# Domesti  s

## Jacqueline Fearn

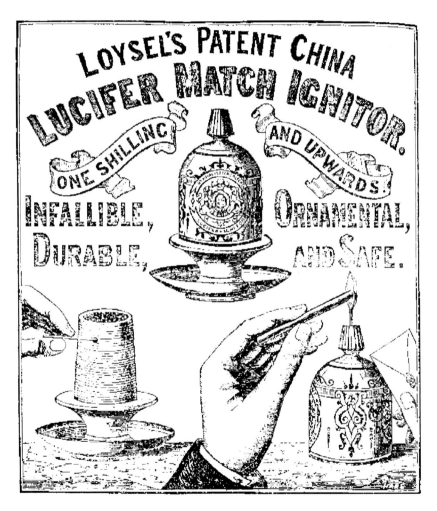

*This delightful item was available in four grades: plain white, with fillets in colour, with fillets of gold, and with artistic designs in colours and gold.*

Shire Publications

# CONTENTS

COVER: *An open hearth, with spit racks above the mantelshelf, and (right) a bread oven. Along the mantelshelf: metal scoop, milk carrier, wooden spoons in a stoneware jar, butter hands, goffering iron, horn beakers, sugar nip, pewter tankard. Centre: pothangers, chimney crane from which are suspended a kettle, another pothanger and a preserving pan. Foreground: a tall trivet, a long toasting fork, a pewter dish, an iron griddle (in front, copper pans), the fireback behind firedogs with a basket spit, fire irons (in front, water jugs and a mortar and pestle). Centre bottom: two trivets.*

ACKNOWLEDGEMENTS

Photographs and other illustrations are acknowledged as follows: York Castle Museum, pages 6, 28 (top left); Geffrye Museum, London, pages 9, 12 (bottom), 17 (top); Holsworthy Museum, page 19 (bottom right); author's collection, photographed by Michael Bass, cover and pages 3 (top), 4 (bottom), 5 (centre), 16 (bottom), 17 (bottom), 18 (bottom left), 20 (bottom), 24 (top). All other illustrations are reproduced by kind permission of the Museum of English Rural Life, University of Reading.

British Library Cataloguing in Publication Data: Fearn, Jacqueline. Domestic bygones. – 2nd ed. – (Shire album; no. 20) 1. Household appliances – England – History 2. England – Social life and customs I. Title 683.8 ISBN 0 7478 0392 7.

Published in 1999 by Shire Publications Ltd, Cromwell House, Church Street, Princes Risborough, Buckinghamshire HP27 9AA, UK. (Website: www.shirebooks.co.uk)

Copyright © 1977 and 1999 by Jacqueline Fearn. First published 1977; reprinted 1980, 1981, 1983, 1985, 1987, 1990, 1991, 1992, 1993 and 1996. Second edition 1999. Shire Album 20. ISBN 0 7478 0392 7.

Jacqueline Fearn is hereby identified as the author of this work in accordance with Section 77 of the Copyright, Designs and Patents Act 1988.

Printed in Great Britain by CIT Printing Services Ltd, Press Buildings, Merlins Bridge, Haverfordwest, Pembrokeshire SA61 1XF.

*Travelling curling irons, which fold into the slim green leather-covered box at the top, the metal arms poised above an oil-fed wick in the brass box, the lid of which folds over. The match box (left) has a flint on which to strike the matches and is covered in maroon leather.*

# INTRODUCTION

Homes nowadays contain a range of basic equipment far more uniform than at any other time since the middle ages. New homes provide a background of built-in features differing only marginally according to taste and expense; the latter half of the twentieth century produced standardisation in domestic requirements and expectations as in other aspects of life. Since the Second World War domestic technology has progressed so quickly and the standard of living has risen so markedly that it is easy to forget that in the early twentieth century the domestic life of the poorer classes in many cases compared more readily with that of their eighteenth-century

*Pattens to keep the feet above the ground were worn in any watery or muddy situation, either with or without shoes, especially in the country.*

3

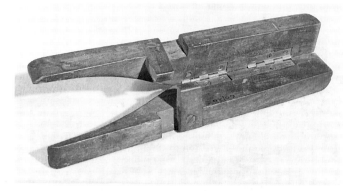

Left: *A folding boot jack. It was opened out, and the protecting pieces were fitted around the heel and pushed down with the handles.*

Right: *A quill pen cutter and case. The round flange at the end (left) was raised and the quill inserted from the end. When the flange was pressed down, the quill was cut to shape. The small knife was used to make a central slit at the nib of the quill.*

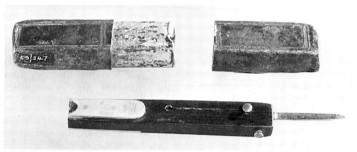

counterparts than with the contemporary middle classes. The transition to the universal use of gas and electric power in the home speeded up in the postwar period so that the younger generation has difficulty in identifying some of the equipment their grandparents, and even their parents, remember and, often, used.

In looking at domestic bygones, then, two factors must be borne in mind. Firstly, at any time up to the twentieth century developments in methods and equipment took a long time to spread through society and sometimes never did. As a result, many interesting items would be found only in upper- and middle-class homes: the lower classes either went without or improvised. Secondly, it is surprising how recently some items we regard as quaint and hardly usable were generally accepted as standard. Much equipment like this, which used to moulder in junk shops, is now being reclaimed and given new status. There are also differences between life in the towns and life in the country and consequently in domestic furnishing.

Looking at households from the middle ages to the present, the progression towards greater comfort and efficiency depended at first on manpower and later increasingly on mechanics as well until modern exploitation of manufactured energy. In all rooms but the kitchen furnishings were at first few, and by our standards cumbersome and uncomfortable, though attractive in their functional simplicity of design. As household life became less communal and houses began to acquire several living rooms, differences in style developed between furnishings and equipment for various rooms.

*A tin sandwich box.*

4

Left: *A stomach warmer, to be filled with hot water.*

Below: *A glass insect trap. Sweet water was poured into the reservoir and left to attract its prey.*

The number of necessary items increased as expectations grew, and the number of furnishings increased as designers and manufacturers determined to supply every need and more. The industrial revolution provided the means of supplying both necessities and former luxuries to a large number of people in new and improved materials. During the latter half of the eighteenth century increased production seemed to go hand in hand with good design and maintained standards of quality; in the nineteenth century increased mechanisation supplied much larger quantities of mass-produced, cheap articles. The expanding middle class followed fashion, for example the Gothic movement of the second half of the century, which, with its solid, profusely ornamented forms, dominated design and helped create the heavily cosy environment so loved by the Victorians. The parallel development and exploitation of improved materials and the irrepressible urge to invent meant that homes were full of aids to comfort and efficiency and every available surface was covered with ornament – the more, the grander.

At the beginning of the twentieth century the clutter began to diminish. The Edwardians reacted against the confusion and, through movements like Art Nouveau and the study of domestic design, forms were simplified and a return to functional design was begun. The second postwar period saw the refining, almost out of existence, of decoration in some aspects of household design. It may be that too much of the tidy 'Scandinavian' look provoked the gradual return of interest in the objects that cluttered Victorian homes!

The greatest opportunity to inspect or acquire lies in nineteenth- and twentieth-century bygones: more items were produced more recently and therefore more survive and, as there was no hesitation by inventors, manufacturers or salesmen to fill the smallest gap in the range of useful devices, the field is vast.

*An ingle fireplace, reproduced in York Castle Museum, with spit racks above holding spits not in use. On the firedogs is another type of basket spit (see cover), together with a dripping tray set in front of the bar grate containing the fire. The spit is turned by the weight-driven spit-jack (right). To the left are a long-handled frying pan and a ladle, to the right a round-bottomed saucepan, a toaster and a hanging griddle.*

# THE HEARTH

The centre of life in the home of the common man until recent times was the kitchen, and the focal point was the hearth. Around the heat of the fire hovered all the activities of the house, and the ingenuity of succeeding generations devised means of using one source of heat for many purposes. Keeping the fire going was a priority (not least because striking a light from the tinder box, if available, was an uncertain and unwelcome chore), and the day's work could hardly be started without it. During the night the embers were covered with a metal dome called a *couvre-feu*, with a gap at the side for a slight draught, and in the morning they were blown into flame with bellows. The major innovation in kitchen design between the twelfth and nineteenth centuries was moving the fire from the centre of the room to the thickened end wall, and into its own fireplace with a chimney (a development not universally adopted until the sixteenth century). The space was built wide and deep to provide a protected area of heat and warmth (and, in many cases, of eddying smoke). The

*A couvre-feu, which was placed over the embers of the fire at night – hence 'curfew', the time to cover the fire and put out the lights.*

fire was in the middle, either on the ground or frequently raised slightly on bricks. The back was usually protected by a cast-iron *fireback*, sometimes attractively moulded. The logs were supported by *firedogs* of many designs: *cup dogs* had small cup baskets on top for holding drinking vessels to warm. Most dogs had hooks at the backs of the uprights where the spit could rest for roasting. Later horizontal bars, laid across between the dogs, developed into *fire-baskets* or *braziers*, or *dog grates* to hold coal, raised above the hearth to provide the necessary updraught. Although a patent for a type of enclosed grate and oven was taken out as early as 1635, most households still used an open fire two hundred years later.

Where there was a *bread oven* it was usually near the fire and often opened into the fireplace. It was usually built in brick with an arched roof. It was heated by filling with brushwood which was set alight and the door was closed. When the fire had burnt out the ashes were raked out, sometimes through a narrow chute in the floor, and loaves and cakes were put in. Batches for baking were inserted and withdrawn on the end of a long wooden paddle called a *peel*.

In larger, affluent houses fireplaces in drawing and dining rooms followed fashion, their shape, size, mantels and fittings changing accordingly, and the trend towards coal fires gradually altered the shape of the grates. Their function was not confined to space heating, and many of the high-quality brass and copper accessories which survive today were not used in the kitchen. Elegantly designed *trivets* to stand in the hearth or hang from the bars of the grate, four-legged *footmen* or stools for the same purpose, toasters and spirit-heated kettles on tall or short stands were all designed for the elegant teatime around the fire. Fire irons multiplied in variety if not efficiency, the shovels giving most scope to the craftsman. A raised grate required something to stop embers rolling beyond the hearth, and in the seventeenth century fenders arrived, and later fireguards. The handy box for keeping spills – lengths of diagonally twisted wood shavings for taking a light from the fire – replaced the brand tongs. When bellows, blowers and scuttles had been embellished, too, there was still scope for metal hearth ornaments whose only purpose was to provide further decoration.

*Rotary fan bellows: some brass drawing-room models were very fine.*

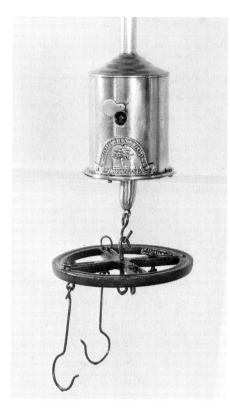

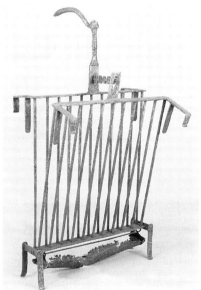

Above left: *A bottle jack with the wheel from which the roast could be suspended by various combinations of long hooks. The jack was wound up by a key inserted in the hole, shown open.*

Above right: *A hastener. The jack was suspended from the bar at the top, the cylinder remaining above the hastener while the wheel was inside, revolving. Progress was inspected through the door.*

Left: *A gridiron. The hooks at the top ends could be attached to fire bars. The juices and fat ran into the shallow container, now very decayed, at the bottom.*

*Wrought-iron chimney crane with three movements. Pots and pans could be raised, lowered, moved along and swung over the hearth. In use, particularly in the south-east of England, between the sixteenth and eighteenth centuries.*

# COOKING AT THE FIRE

Meat was cooked, then as now, by roasting, broiling, frying and stewing. The important element in roasting was the steady rotation of the joint in front of the fire, which had to be 'bright'. Until the eighteenth century a *horizontal spit* was commonly used, a metal rod that rested on hooks or notches in front of the *andirons* (firedogs with spit hooks), with a handle at one end. Smaller joints or game were enclosed in a *basket* or *cradle* spit. *Wheel* spits were devised which were turned by a rope running to a pulley wheel. Turning power was transmitted, by way of further ropes and wheels, at first by hand, by specially devised dog *treadmills*, set in the wall, and eventually by gravity relying on the slow descent of a winched weight, and by *smoke jacks* within the chimney operated by the hot air of the fire. As the meat roasted the fat dripped into an oval iron dish set beneath to catch the 'dripping'.

By the end of the eighteenth century a spring-driven mechanical jack, called a *bottle jack*, came into use. The roast was suspended from a wheel under the bottle-shaped brass cylinder, which contained a clockwork spring. The roast would revolve for an hour or so as the spring unwound. This device was held by a clamp screwed to the mantelpiece above the fire. It was realised that reflected and contained heat speeded up the roasting process, so semicircular sheet-metal *hasteners*, burnished bright, were placed facing the fire, half surrounding the revolving joint. This idea was developed to produce the standard hastener or *Dutch oven*.

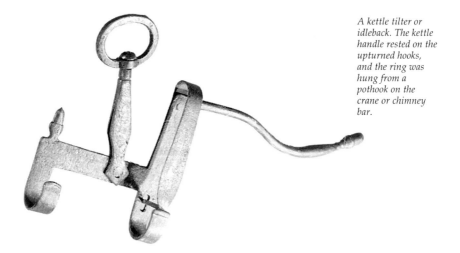

A kettle tilter or idleback. The kettle handle rested on the upturned hooks, and the ring was hung from a pothook on the crane or chimney bar.

Small pieces of meat or fish could also be roasted or broiled (grilled) on a *gridiron*, constructed with hollow bars down which the juices ran to a container at the end. These were of many designs, including the hanging grill that enclosed the meat in a slim metal cage, in use over a long period. There were many variations and improvisations to bring food of all sizes close to the fire.

From the bronze age to the end of the nineteenth century (in some places) the important business of boiling water was done over the fire in a large metal pot or *cauldron*. Ultimately cast in iron, these vessels had rounded bottoms, to encourage all-round distribution of heat, and often three little legs. It was clearly advantageous to suspend pots over the fire, and systems of chimney hooks, bars, sways and cranes developed which were standard equipment until the nineteenth century. The most elaborately versatile of these was the *chimney crane*, a wrought iron fixture incorporating a variety of devices (and ornamentation). At first a simple, pivoting, right-angle bracket fixed on the back wall of the fireplace, this device was adapted to move in three planes so that the housewife could use all parts of the fire at the heat she required. *Pot*

A fine eighteenth-century bell-metal skillet with WASBROUGH, the name of the manufacturer, boldly cast into the handle.

10

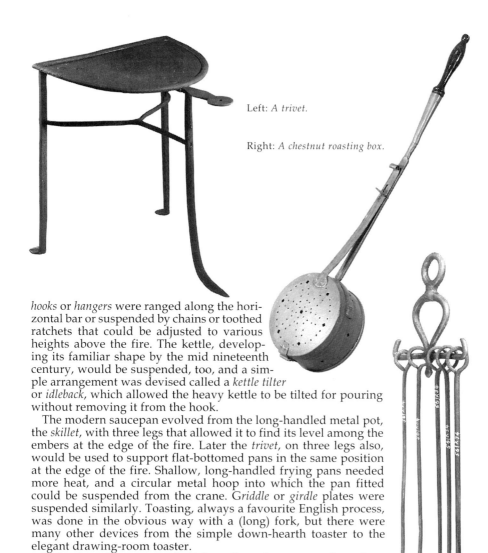

Left: *A trivet.*

Right: *A chestnut roasting box.*

*hooks* or *hangers* were ranged along the horizontal bar or suspended by chains or toothed ratchets that could be adjusted to various heights above the fire. The kettle, developing its familiar shape by the mid nineteenth century, would be suspended, too, and a simple arrangement was devised called a *kettle tilter* or *idleback*, which allowed the heavy kettle to be tilted for pouring without removing it from the hook.

The modern saucepan evolved from the long-handled metal pot, the *skillet*, with three legs that allowed it to find its level among the embers at the edge of the fire. Later the *trivet*, on three legs also, would be used to support flat-bottomed pans in the same position at the edge of the fire. Shallow, long-handled frying pans needed more heat, and a circular metal hoop into which the pan fitted could be suspended from the crane. *Griddle* or *girdle* plates were suspended similarly. Toasting, always a favourite English process, was done in the obvious way with a (long) fork, but there were many other devices from the simple down-hearth toaster to the elegant drawing-room toaster.

Before bread ovens were built in ordinary houses, or where there was none anyway, the cauldron was sometimes inverted on a metal plate on a slow fire to serve as an oven. There were several variations on this idea, and Gertrude Jekyll's *Old English Country Life*, published in 1925, shows a Yorkshire kitchen about 1900 where a cake is baking in a suspended, squarish pot that has glowing peat on the lid. This was uncommon so late: by that time the majority of households had some form of *kitchen range*.

It had never been easy to carry on all necessary operations at an

Right: *Skewers on their hook, obviously designed to secure large pieces of meat.*

11

Wrought-iron toasters of the seventeenth and eighteenth centuries, the smaller for toasting bread on a floor-level or 'down' hearth, and the elegant standing model for use in front of the kitchen range or the drawing-room fire, with a plate carrier below. Both models revolved.

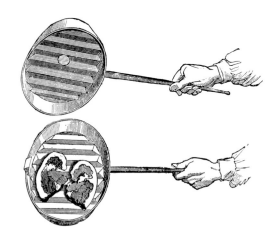

This cast-iron 'Batchelor's frying pan' must have been rather heavy. It was featured in Warne's 'Model Cookery and Housekeeping Book' at about the end of the nineteenth century. This book, designed for modest households and very much smaller than Mrs Beeton's 'Book of Household Management', covers most aspects of day-to-day housekeeping, including illustrations and descriptions of kitchen equipment.

# LEAMINGTON KITCHENER

## WITH PATENT REGULATOR.

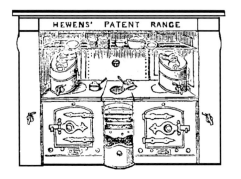

*An example of a kitchen range. This one is advertised on the basis of fuel saving by the 'patent regulator' and comes from Warne's 'Model Cookery and Housekeeping Book'. One oven is ventilated for roasting, the other is closed for bread and pastry baking. The brass taps at the side were for drawing off hot water from the back boiler. This type of range was recommended by Mrs Beeton and in 1859-61 cost from £5 15s to £23 10s.*

open fire: more than one fireplace might be needed and in major establishments three or four. At the lowest end of the scale, the main baking of a poor household without an oven had to be sent out to the local baker, who also bake-roasted their meat for them. In areas where coal was available, by the early eighteenth century the fire was usually contained in a grate secured to the back of the fireplace, and soon after grates flanked by iron boxes called *hobs* appeared, enabling food to be kept hot or even a kettle to be boiled. At the same time manufacturers began to produce cast-iron ovens with their own grates that could be built into the side of the fireplace. The first man to patent a design that moved the oven next to the fire, so forming a kitchen range, was Thomas Robinson in 1780, but other innovations soon followed: a water boiler on one side or at the back, from which water was drawn off by tap; flue systems around the oven; brick-lined ovens heated by the fire; detachable trivets, grills and grids. The culmination of all this was the completely closed range, clean and much more fuel-efficient. Gradually some form of range began to appear in most types of home as the use of coal became more widespread during the nineteenth century. Most of the old implements and equipment could be used with the open range, though scaled down and lighter, like the copper or steel pans already in use. It was still a very busy scene for some time, with trivets over and off the fire or cranked beside the bars; the bottle jack and Dutch oven were still in use, for meat was still open-roasted for preference; the kettle tilter was still a better option than lifting a heavy kettle. Add to these the chestnut-roasting boxes, warming pans, ale warmers and the impedimenta for feeding and maintaining the fire, all of which surrounded the fireplace. Thank goodness the temperamental damper systems, emery paper and blacklead to keep the range from rusting are only memories or curiosities!

# PREPARING FOOD

In country households food preparation started beyond the kitchen door, the housewife doing or supervising many of the operations later taken over by the dairyman, butcher and others: in towns these trades, with subdivisions such as cheesemonger and poulterer, were established early. In the kitchen itself and its allied larders there was still a great deal to do, both planning and processing, while food had to be kept without refrigeration.

As with cooking at the fire, most of the utensils used from early times to prepare food look very similar today. Wooden bowls, bins, jugs, platters and implements were joined by pewter, glazed pottery and tin: mass-produced china predominated from the nineteenth century, to be challenged in recent times by plastic. Kitchen knives have changed very little in shape, while skewers are more refined and uniform; meat forks have diminished in size and length since there is no need to lift large hunks of meat from a deep pot over a hot fire; spoons and skimmers have contracted too. Stainless steel is now universally used except in the case of the sharpest knives.

Preserving meat was a major problem both in the long term, for the winter months, and during shorter periods of hot weather. After the autumn slaughter meat was usually dry-salted in barrels, with the addition, in time, of potassium nitrate as a drying agent. Brine salting, in troughs or barrels, followed by hanging, could be employed for long-term or short-term storage, depending on how long the meat was left to steep. Goat and mutton were salted as well as beef, and pork was the most palatable at the end. Head and trotters could be potted into brawn, sealed with an acid liquor in glazed earthenware and later with butter. From the seventeenth century, when breeding stock began to be improved and winter feeding on turnips started, there was less need for slaughter, but people continued to do some salting and pickling (in vinegar solutions and wine at best, salt and water in general).

The flavour of preserved meat had to be improved before serving, and the condition of bought meat, especially in the towns, was often poor. This accounts in part for the great range of herbs and spices employed in cooking all kinds of food: poor flavour needed to be disguised, and the palate came to expect heavy spicing. Most households grew their own herbs, and the initially expensive imported spices were carefully stored in secure *spice boxes*, under lock and key for larger supplies. Herbs were pounded or ground fine in the *mortar*, which came in several different sizes. Spices were treated similarly or, as with pepper, milled to fineness, or steeped in liquor; nutmeg, important in many dishes and in drinks, was grated (pocket graters were available for those who did not wish to risk going without it).

Left: *A tin spice box with named containers for six spices.*

Above left: *A well-used herb chopper.*

Above right: *A beautifully turned wooden spice box.*

Left: *A salt box that hung on the wall in a dry place, probably in or near the fireplace.*

*A large mortar (12 inches, 30 cm across) and pestle.*

Sugar, brought into use by the returning Crusaders but known for longer, was being used in increasing quantities by the noblest households by the thirteenth century. At first the producers refined it, and at the end of the process sugar liquid was poured into moulds that hardened into conical loaves weighing between 3 and 14 pounds (1.4–6.4 kg). For domestic use pieces of the loaf were chipped off and pounded. It was regarded at first as both a spice and a medicine (especially for colds) and was not seen simply as a sweetener; it was treated as another flavour. Eventually it contributed to the decline of true spices when used in greater quantities from the mid seventeenth century, when England gained her own sugar colonies. It was not cheap and was taxed until the nineteenth century, so large quantities of honey were used. Yet by this time 12 pounds (5.4 kg) of sugar per head of the population were consumed annually. Several grades were produced.

It was not possible to make the light confections we know today as pastries,

*Steel pint and half-pint milk carriers with brass handles and loops, and a light-alloy half-pint milk measure.*

Left: *Sugar loaf and cutters, used well into the nineteenth century.*

Below: *A piggin, used either as a drinking vessel or as a ladle, more often in the dairy.*

puddings and cakes until the rising effect of beaten eggs was realised, granular sugar was available and flour was finely milled. For example, gingerbread was originally an uncooked cake made with compressed breadcrumbs bound with honey and flavoured with liquorice and other spices. Printed moulds pressed out thin cakes for drying off in a cool oven. Eventually ginger predominated, flour began to replace breadcrumbs and the cakes were baked. Small cakes – sometimes like simnels preboiled – were baked from medieval times, but larger cakes were really a form of highly enriched bread, leavened by ale-barm: it was not until the eighteenth century that eggs were trusted to support a large cake.

*Wooden utensils. (Left to right): a cabbage or spinach press for squeezing water out of the vegetables; a potato masher; a butter worker, for squeezing the water from freshly churned butter; butter hands or pats for dividing up and shaping the butter. Bottom: an oatmeal crusher.*

Far left: *A beautifully made bentwood flour barrel.*

Left: *A salamander, which was heated in the fire and then used to brown meat and pastry that were cooked but still unappetisingly pale. The iron shaft was attached to a longer wooden handle; otherwise it would be too hot to grasp after it was drawn from the fire.*

Right: *A gingerbread mould at the top flanked on the left by a biscuit cutter and pricker combined, on the right by a biscuit pricker. Below them are five biscuit stamps.*

Left: *(Top to bottom) an icing or biscuit gun; an apple corer made from a sheep bone; a metal lemon squeezer.*

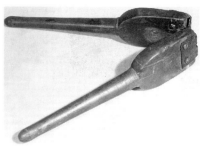

*A wooden lemon squeezer.*

18

*Gingerbread blocks, all 5 inches (13 cm) wide, representing Wellington perhaps, a wagon and Nelson's coach.*

Biscuits have a long history, from the 'thrice-baked bread', ship's biscuit and the batter-based wafer down to our own home-made biscuits, shaped by moulding or cutting. Hot ovens are used today: then 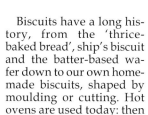 it was the bread oven after the bread was removed, or an additional cooler oven set in the fireplace. Pastry also needs a hot oven, and pies were often browned with a *salamander* or *pastry browner* (also used to brown spit-roast meat).

Jelly was made from early times from bonestock and calves' feet, most often for savoury dishes but sometimes as a sweet with a flavouring of honey, sugar, wine or fruit puree. The potential for shaping was realised early, and the preoccupation was with colour, not flavour. Our wealthier ancestors liked their food to look beautiful. When tasteless isinglass or gelatine began to be commercially produced (from bones and hides or, the best, from the swimming bladders of sturgeons) in the sixteenth century and when aspic

Above: *A tin jelly mould.*

Left: *A knife machine, showing slots for three knives, by Spong, a name famous for kitchen machines until very recently. The cleaning powder was poured in through the hole with the round stopper on the right-hand side. The strops inside revolved when the handle was turned. As well as the guarantee (above the handle), the machine still has the instructions pasted to the left.*

19

An apple peeler. On this model the apple is held on the fork at the top and the knife unit swings up to rest just above the central bar. When the handle is turned the screw moves the knife unit along. The contours of the apple are approximately accommodated by spring action.

was concocted in the eighteenth there was an increase in the scope and use of moulds of pottery and, in numerous designs, of brass, copper and tin.

The expanding urban middle class of the nineteenth century provided a market avid for gadgets. Some developments like mincers and grinders were excellent and are little changed today, even if electrically powered. Others have been displaced by food manufacturers – few people roast their own coffee beans or mill their own flour – and many, predictably, saved neither time nor effort once out of the salesman's hands and so passed into oblivion. It is sometimes difficult to judge how efficient the device might be deemed: quite a number of knife-cleaning machines have survived, and this suggests that many were sold. The established method of cleaning steel knives was by stropping on a board faced with leather or rubber compound in conjunction with emery powder. Although three or more knives could be cleaned simultaneously in a machine, it still appears that the board was not superseded over seventy years: perhaps only larger establishments profited by the invention.

Pressure cooking was experimented with as early as 1682, when a model was demonstrated by its results for the Royal Society. With a pressure of 35-50 pounds a square inch, the 'New

A vertical marmalade cutter with a screw clamp to attach it to the kitchen table. The double-edged knife cuts through the fruit as it is pressed through the feeder tube and out to a bowl on the other side.

*The patent Digester (centre), forerunner of today's pressure cooker and used to make stock, flanked by a Saucepan Digester (left) and a Stewpan Digester (right). Simmering time was eight to ten hours. These innovative items were featured in Warne's 'Model Cookery and Housekeeping Book'.*

Digester for softening bones' was highly dangerous, if impressive. The idea was not developed until the nineteenth century, when the digester was suitably equipped with a steam release-valve, raised when pressure exceeded 3 pounds, which meant that the process took rather longer than in the modern 15 pound cooker. It was highly regarded.

Only relatively recently was breadmaking abandoned in many country areas. Baking for the week was a large undertaking, and several different kinds of bread might be baked. The volume of dough can be judged from the size of the dough trough in which it was left to rise. But the chore from which one feels most remote is lard-beating, when the fat collected from the simmered lard jar was beaten, to expel water, producing a cheap fat that shortened the pastry and fried without spitting.

*A dough trough or bin, now without its lid, where the dough could be left to prove.*

*A cheese cradle.*

# EATING AND DRINKING

Until the sixteenth century most tables were laid with wooden utensils – platters. bowls, spoons and beakers (which might also be of horn). Pewter utensils and salts were used by the middle classes, and the nobility used silver also and sometimes gold. Until the fifteenth century and sometimes afterwards meat was served on to large trimmed squares of coarse bread – at least four days old – which served as plates and mopped up the gravy. They were called *trenchers*, and as they were joined and replaced by wooden plates the name was retained. Several shapes of trencher developed, and they survived in country areas as late as the twentieth century even though from the seventeenth century first pewter and tin, then glazed pottery and china plates came in.

Knives were often carried as a matter of course; cutting meat at table (and lifting food to the mouth) was only part of their use. In the seventeenth century English prejudice against forks receded, and the introduction of a two-pronged fork began a movement among the genteel towards special implements – knives with rounded ends and spoons with rounded bowls and flat handles in pewter or tin: the traveller carried his own set. Canteens of cutlery appeared in the eighteenth century, and tableware multiplied in the nineteenth with special sets for different foods, napkin rings, condiment sets, egg toppers, cheese cradles and cheese pickers. The delightful moustache cup and saucer illustrate the enthusiasm to supply an elegant solution to every problem, real or imagined.

Hot drinks like chocolate, coffee and tea had required a change in drinking vessels. Pewter and tin were not only uncomfortably hot but very often imparted a metallic taste, if nothing worse, to the drink. Hence came a major stimulus to the

22

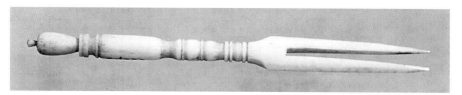

*A cheese picker for transferring a piece of cheese from the cradle to the plate.*

*An elegant moustache cup and saucer.*

*A tin boot (or slipper) ale-warmer, which was pushed close to the fire.*

*A nineteenth-century egg-shaped mineral-water bottle, which would have had a cork stopper.*

china and pottery industries of the new industrial age to produce glazed, tasteless cups. Wine vessels, again metallic, began to be produced in glass from the seventeenth century, partly for the same reason, although the palate must have been accustomed to a certain amount of taint.

Wine bottles had developed to their cylindrical shape (most suitable for binning horizontally) by 1750. This was the end of a progression from the invention of the corkscrew, allowing tight corks to be extracted from the bottleneck, and the development of bottle cork-pressers to insert the cork securely. The problem of keeping beer in bottles without further fermentation was not solved until the invention of the screw stopper in 1875, but from 1814 mineral water was successfully bottled using cork stoppers in egg-shaped bottles that allowed the cork to remain wet and tight because the bottles had to be stored horizontally.

*A travelling knife and fork set with its leather case.*

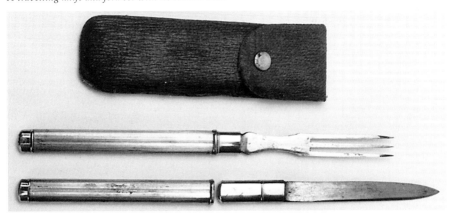

# LIGHTING

One of the earliest forms of lighting, known to the Greeks and Romans, was the *rushlight*, still in use in the latter half of the nineteenth century. At the height of summer all but the able-bodied men would gather the longest and fattest common rushes; these were then steeped in water to prepare them for peeling. The rush had to be stripped to the pith while leaving a narrow strip of skin running from top to bottom to support it. These wicks were laid out to weather, bleach and dry in the sun and then were gathered into bunches, for dipping at home or for sale. According to Gilbert White of Selborne, writing in 1775, one pound of rush-dips, costing one shilling, could contain 1600 dry rushes.

The grease for dipping was usually bacon or, better, mutton fat skimmed from the cooking pot. This was melted in a long oval cast-iron *greasepan*. Rushes were laid in the pan until saturated, then taken out and allowed to dry; they were sometimes redipped to add another coat of fat. Such lights might burn for half an hour and, again according to White, even if made with bought rushes and grease at fourpence a pound, would give five and a half hours of light for a farthing.

The long thin taper produced, too fragile to be held upright, was supported in the middle between iron jaws or *nips*, one of which was long-spiked into a wooden block, in the simplest design. There were many variations in size and design of holders but the taper had always to be firmly held by a riveted, counter-weighted or spring-held closure. When the flame burned down, the taper would be moved up until only a small stub was left, or both ends could be lit – extravagantly burning the candle at both ends.

Rushlight holders are often found with fittings for candles as well; shorter rush candles were used, made in the same way as rushlights but dipped several times, and then various other forms of tallow candle. From the fourteenth century candle-moulding had produced a firmer, regularly shaped candle more efficiently. Wooden or metal *candle stools*, producing a pound or two of candles of uniform or different sizes, were used either by travelling candlemakers or the women of the household. Candles were produced commercially in large quantities, increasing as time went on, but not cheaply as it was a fiddly process.

The light produced was dim and erratic, streams of grease poured down the holder into the sconces, and the loosely twisted wicks, burning more slowly than the grease, had to be trimmed before they collapsed (and the smell was unpleasant). Wax candles gave a better light but were very expensive as they were made from rolled sheets of wax. In the eighteenth century quickly vaporising, clean-burning spermaceti from the whale came into use; early in the nineteenth, hardeners, plaited cotton wicks and finally paraffin wax superseded the fats, producing the firm long-burning candle with a clear and almost smokeless light we still use.

*A simple rushlight holder. The hinged arm serves also as a candle holder. The rushlight would be held diagonally and moved as the flame moved down the light.*

*A wooden candle-making battery with metal moulds and a wax pourer. The wick was threaded through the mould, knotted at the tapering, almost sealed end and secured at the top by tying on to wires that were fixed horizontally across the stool in line with the tops of each row of moulds. The tallow was then poured in.*

Right: *A selection of candle snuffers with a candle trimmer on the left.*

*A tinderbox with flint and steel. Dry rags were kept in the box so that a spark would set them smouldering to be blown into flame. Many boxes incorporated a candle holder so that the candle could be lit and the flame retained.*

26

*Candles were stored in wall hanging boxes like this.*

By the nineteenth century there was competition from developments in the oldest form of lighting known – the *oil lamp*. In its simplest form all that was needed was some kind of wick to burn in a container of oil. Early English lamps were of crude earthenware or iron with a spout at the side of the dish to make a lip for the wick: few survive. *Crusie* lamps were of Celtic origin, dating from the early iron age, burning fish oil as fuel and continuing in use in some remote areas until modern times.

There were many developments of the principle, using different oils, until in 1784 Argand, a Swiss physicist, invented a round burner for colza or vegetable oil, with a tubular wick and iron chimney; his partner, Quinquet, invented a glass chimney, and in 1800 Carcel invented a clockwork pump for raising and feeding in the oil. The Moderator lamp of 1836 had a spring-loaded piston to feed fuel under pressure. When clean, easy-burning paraffin came in 1847, developments were confined to paraffin lamps, and the profusion of lamps and heaters of every size and shape dates from this time.

*The small container (right) of this crusie would be filled with oil and placed on the hook on the upright of the large container.*

*A dolly tub and three-legged dolly stick or peg in use.*

# WASHING

The weekly wash is a modern phenomenon: our forebears did not share our pre-occupation with cleanliness, partly because they did not see the need and partly because it was a lot of trouble. The poor did not have a great deal to wash – their woollen items washed badly – and, except in the specially equipped houses of the very wealthy, the process of washing, beating, rinsing and laying out to dry was done outside. Even when materials like linen and, later, cotton became more widespread there were few households between the sixteenth and nineteenth centuries where a wash was undertaken more than once a quarter.

The great households set the pattern, and by the sixteenth century they had premises for washing where water could be heated and clothes washed and laid out

*Washing bats, which pounded the clothes.*

28

*A box mangle. The box was filled with stones and travelled backwards and forwards over the rollers, around which the clothes were wound. At the end of its progress in each direction it tipped to allow the rollers to be removed. Turning the wheel was hard work because the box was so heavy.*

to dry, starched, ironed and lavendered – the laundry. 'Whitsters' or laundry maids could be permanently employed even though their work was intermittent, or, more often, washerwomen came for the day to 'buck' the clothes, a word describing the whole process or any part of it.

Most often using *lye*, a carefully concocted mixture of ashes, boiling water and slaked quicklime (soap was expensive until the nineteenth century and often, then, home-made), the clothes would be buffeted with bats against stone or board, rinsed and *mangled* by pressing between bats and rollers. By the late eighteenth century a *box mangle* might be used. Linens and cottons were laid out to bleach in the sun; woollens could be stretched, to prevent shrinking, from *tenterhooks* on the crossbar between two tall upright poles, and weighted at the bottom with a heavy object.

*A linen press used to press heavy fabrics into their creases.*

29

*A goffering machine. The hollow ribbed brass rollers were heated by the insertion of hot circular irons.*

In the nineteenth century a washerwoman in the towns often collected laundry to do at home: if she had one, she might also hire out the use of her mangle. Her facilities were usually limited to the deep tub (eventually galvanised, with a *dolly* to plunge and agitate the clothes or a ribbed *washing board*) then in general use. Increasingly a small wash-house was built adjoining the house; in it there was a copper boiler with a fire underneath, which was superseded only by the modern washing machine.

The first washing machine, by Sidgier, appeared as early as 1782. It was a rotating drum operated by a geared handle: the first manufacturer of any quantity was Thomas Bradford in the 1860s.

'Dashing away with the smoothing iron' (a little rhyme which neatly illustrates the bother of washing) required a good fire and more than one iron. The *flat iron* was heated by placing near the fire or on the hob. It was the simplest and most usually found and, as with the other, hollow types (into which hot slugs of metal or glowing charcoal were inserted), it required a nice judgement in use. The mid sixteenth century brought its own special trial to the laundry when starched ruffs came into fashion. The material was damped with starch solution, then folded successively over a wooden *poking stick* of the required size until the band was crimped.

More efficient devices followed quickly. A hollow metal poking stick on a stand was heated by the insertion of a hot metal rod so the material dried as it was pleated. This was called a *goffering stick* or *iron*. The *goffering stack* was another development – the damp material was passed in and out of wooden slats which were weighted down by the secured top bar. The process was finally rationalised in the *crimping machine*.

*A goffering stack used for cold crimping. Damp linen was threaded through slats, which were held down by the top bar and secured by wedges.*

# FURTHER READING

Caspall, John. *Fire and Light in the Home pre 1820.* Antique Collectors' Club, 1987.
Davies, Jennifer. *The Victorian Kitchen.* BBC Books, 1989.
Eveleigh, David J. *Brass and Brassware.* Shire, 1995.
Eveleigh, David J. *Candle Lighting.* Shire, 1985; reprinted 1995.
Eveleigh, David J. *Firegrates and Kitchen Ranges.* Shire, 1983; reprinted 1996.
Eveleigh, David J. *Old Cooking Utensils.* Shire, 1986; reprinted 1997.
Fearn, Jacqueline. *Cast Iron.* Shire, 1990.
Gentle, Rupert, and Feild, Rachel. *Domestic Metalwork 1640-1820.* Antique Collectors' Club, 1994.
Ingram, Arthur. *Dairying Bygones.* Shire, second edition 1997.
Lindsay, J. S. *Iron and Brass Implements of the English House.* Tiranti, 1970.
Meadows, Cecil A. *Discovering Oil Lamps.* Shire, second edition 1987; reprinted 1995.
Perry, Evan. *Collecting Antique Metalware.* Country Life, 1974.
Pinto, Edward H. *Treen and Other Wooden Bygones.* Bell, 1969; reprinted.
Sambrook, Pamela. *Laundry Bygones.* Shire, 1983; reprinted 1997.
Wilson, C. Anne. *Food and Drink in Britain.* Peregrine, 1976.

# PLACES TO VISIT

Domestic bygones can be seen in many places and the list below gives a selection of some of the best collections to be found in museums. stately homes, farm and industrial museums. Intending visitors are advised to find out opening times and check that relevant items are on show before making a special journey.

*Abbey House Museum*, Abbey Road, Kirkstall, Leeds LS5 3EH. Telephone: 0113-275 5821. (Closed for refurbishment until 2000.)
*Angus Folk Museum*, Kirkwynd Cottages, Glamis, Angus DD8 1RT. Telephone: 01307 840288.
*Anne of Cleves House Museum*, 52 Southover High Street, Lewes, East Sussex BN7 1JA. Telephone: 01273 474610.
*Arlington Mill Museum*, Bibury, Cirencester, Gloucestershire GL7 5NL. Telephone: 01285 740368.
*Beamish, The North of England Open Air Museum*, Beamish, County Durham DH9 0RG. Telephone: 01207 231811.
*Beck Isle Museum of Rural Life*, Bridge Street, Pickering, North Yorkshire YO18 8DU. Telephone: 01751 473653.
*Birmingham Museum and Art Gallery*, Chamberlain Square, Birmingham B3 3DH. Telephone: 0121-303 2834. (Pinto Collection of wooden bygones.)
*Blaise Castle House Museum*, Henbury Road, Henbury, Bristol BS10 7QS. Telephone: 0117-950 6789.
*Bolling Hall*, Bowling Hall Road, Bradford, West Yorkshire BD4 7LP. Telephone: 01274 723057.
*Brighton Museum and Art Gallery*, Church Street, Brighton, East Sussex BN1 1UE. Telephone: 01273 290900.
*Cambridge and County Folk Museum*, 2/3 Castle Street, Cambridge CB3 0AQ. Telephone: 01223 355159.
*Clifton Park Museum*, Clifton Lane, Rotherham, South Yorkshire S65 2AA. Telephone: 01709 382121 (extension 3635) or 823635.
*Cogges Manor Farm Museum*, Church Lane, Witney, Oxfordshire OX8 6LA. Telephone: 01993 772602.
*Cookworthy Museum*, The Old Grammar School, 108 Fore Street, Kingsbridge, South Devon TQ7 1AW. Telephone: 01548 853235.
*Cotswold Countryside Collection*, Northleach, Gloucestershire GL54 3JH. Telephone: 01451 860715.
*Devizes Museum*, 41 Long Street, Devizes, Wiltshire SN10 1NS. Telephone: 01380 727369.
*Erddig*, Wrexham LL13 0YT. Telephone: 01978 355314.
*Fife Folk Museum*, The Weigh House, High Street, Ceres, Cupar, Fife KY15 5NF. Telephone: 01334 828180.
*Ford Green Hall*, Ford Green Road, Smallthorne, Stoke-on-Trent, Staffordshire ST6 1NG. Telephone: 01782 534771.
*Geffrye Museum*, Kingsland Road, London E2 8EA. Telephone: 0171-739 9893.
*Georgian House*, 7 Great George Street, Bristol BS1 5RR. Telephone: 0117-921 1362.
*Georgian House*, 7 Charlotte Square, Edinburgh EH2 4DR. Telephone: 0131-226 3318
*Gladstone Court*, Biggar, Lanarkshire ML12 6DT. Telephone: 01899 21050.
*Gloucester Folk Museum*, 99/103 Westgate Street, Gloucester GL1 2PG. Telephone: 01452 526467.
*Guernsey Folk Museum*, Saumarez Park, Castel, Guernsey, Channel Islands GY5 7UJ. Telephone:

01481 55384.
*Gwent Rural Life Museum*, The Malt Barn, New Market Street, Usk, Monmouthshire NP5 1AU. Telephone: 01291 673777.
*Hall ith Wood Museum*, Green Way, off Crompton Way, Bolton, Lancashire BL1 8UA. Telephone: 01204 301159.
*Hereford and Worcester County Museum*, Hartlebury Castle, Hartlebury, Kidderminster, Worcestershire DY11 7XZ. Telephone: 01299 250416.
*Highland Folk Museum*, Duke Street, Kingussie, Inverness-shire PH21 1JG. Telephone: 01540 661307.
*Mary Arden's House*, Wilmcote, Stratford-upon-Avon, Warwickshire CV37 9YA. Telephone: 01789 298365.
*Milton Keynes Museum*, Stacey Hill Farm, Southern Way, Wolverton, Milton Keynes MK12 5EJ. Telephone: 01908 316222.
*Museum of East Anglian Life*, Stowmarket, Suffolk IP14 1DL. Telephone: 01449 612229.
*Museum of English Rural Life*, The University, Whiteknights, Reading RG6 6AG. Telephone: 0118-931 8660.
*Museum of Iron and Darby Furnace*, Coalbrookdale, Telford, Shropshire. Telephone: 01952 433418.
*Museum of Lakeland Life and Industry*, Abbot Hall, Kendal, Cumbria LA9 5AL. Telephone: 01539 722464.
*Museum of Lincolnshire Life*, The Old Barracks, Burton Road, Lincoln LN1 3LY. Telephone: 01522 528448.
*Museum of Local Life*, Friar Street, Worcester WR1 2NA. Telephone: 01905 722349.
*Museum of Welsh Life*, St Fagans, Cardiff CF5 6XB. Telephone: 01222 569441.
*Number 1 Royal Crescent*, Bath BA1 2LR. Telephone: 01225 428126.
*Old House Museum*, Cunningham Place, Bakewell, Derbyshire DE45 1DD. Telephone: 01629 813165.
*Ordsall Hall Museum*, Ordsall Lane, Salford M5 3AN. Telephone: 0161-872 0251.
*Oxfordshire County Museum*, Fletcher's House, Park Street, Woodstock, Oxfordshire OX20 1SN. Telephone: 01993 811456.
*Portsmouth City Museum*, Museum Road, Old Portsmouth, Hampshire PO1 2LJ. Telephone: 01705 827261.
*Powysland Museum and Montgomery Canal Centre*, The Canal Wharf, Welshpool, Powys SY21 7AQ. Telephone: 01938 554656.
*The Priest House*, West Hoathly, near East Grinstead, West Sussex RH19 4PP. Telephone: 01342 810479.
*Red House Museum*, Quay Road, Christchurch, Dorset BH23 1BU. Telephone: 01202 482860.
*Rural Life Centre*, Old Kiln Museum, Reeds Road, Tilford, Farnham, Surrey GU10 2DL. Telephone: 01252 795571 or 792300.
*Rutland County Museum*, Catmos Street, Oakham, Rutland LE15 6HW. Telephone: 01572 723654.
*Ryedale Folk Museum*, Hutton-le-Hole, North Yorkshire YO6 6UA. Telephone: 01751 417367.
*St John's House*, St John's, Warwick CV34 4NF. Telephone: 01926 412021.
*Science Museum*, Exhibition Road, South Kensington, London SW7 2DD. Telephone: 0171-938 8000.
*Scolton Manor Museum*, Clarbeston Road, Haverfordwest, Pembrokeshire SA62 5QL. Telephone: 01437 731328.
*Shibden Hall*, Listers Road, Halifax, West Yorkshire HX3 6XG. Telephone: 01422 352246.
*Somerset Rural Life Museum*, Abbey Farm, Chilkwell Street, Glastonbury, Somerset BA6 8DB. Telephone: 01458 831197.
*Staffordshire County Museum*, Shugborough, Stafford ST17 0XB. Telephone: 01889 881388.
*The Stewartry Museum*, St Mary Street, Kirkcudbright DG6 4AQ. Telephone: 01557 331643.
*Strangers' Hall Museum*, Charing Cross, Norwich NR2 4AL. Telephone: 01603 667229.
*Sulgrave Manor*, Banbury, Oxfordshire OX17 2SD. Telephone: 01295 760205.
*Towner Art Gallery and Local History Museum*, High Street, Old Town, Eastbourne, East Sussex BN23 8BB. Telephone: 01323 411688 or 417961.
*Tudor House Museum*, Bugle Street, Southampton SO14 2AD. Telephone: 01703 635904.
*Ulster Folk and Transport Museum*, Cultra, Holywood, County Down BT18 0EU. Telephone: 01232 428428.
*Upminster Tithe Barn Agricultural and Folk Museum*, Hall Lane, Upminster, Essex RM14 1AU. Telephone: 01708 447535.
*Vale and Downland Museum Centre*, The Old Surgery, Church Street, Wantage, Oxfordshire OX12 8BL. Telephone: 01235 771447.
*Wayside Museum*, Zennor, St Ives, Cornwall TR26 3DA. Telephone: 01736 796945.
*West Highland Museum*, Cameron Square, Fort William, Inverness-shire PH33 6AJ. Telephone: 01397 702169.
*Weston Park*, Weston under Lizard, near Shifnal, Shropshire TF11 8LE. Telephone: 01952 850207.
*York Castle Museum*, Eye of York, York YO1 9RY. Telephone: 01904 653611.